Communism and the Protection of the Environment

Ideology, Economics & Politics

Brian Meadows

Revive Publications

*A catalogue record for this book is available
from the British Library*

ISBN: 978-1-907962-14-1

Published by Revive Publications

Reading, England

For James

Contents

Preface

The global environmental problems that currently exist have arisen because of the activities of democratic states. In other words, if the environmental impacts of democratic states had never occurred then the human perturbations of the environment would not be large enough to constitute a set of global environmental problems.

If democratic states had never existed would we today face a number of global environmental problems? This is a hard question to answer. In order to make some progress in answering this question I consider the environmental impacts which occurred in the Eastern European communist countries. There was severe environmental degradation in certain parts of these countries; was this degradation due to communism? In an attempt to understand why this degradation occurred I consider the underlying relationship that exists between communist ideology/economics/politics and environmental impacts. We will see that there are a great number of factors which come into play when we attempt to identify the nature of the relationship between communism and the environment.

Introduction

Isolating the role that a lack of democracy played in contributing to the severe environmental degradation in the communist countries of Eastern Europe is a complex matter. Attempting to do this requires an analysis of differing temporal and spatial scales. In *Chapter One* the spatial scale of the Eastern European countries themselves is utilised to examine the particular environmental problems that they faced. Then, in *Chapter Two,* the wider scale of the 'Second World' is considered to appreciate the environmental impacts of the underlying communist framework. A wide temporal scale is then adopted in

Chapter Three to examine how the severe environmental degradation that occurred when the West industrialised compares to the degradation in Eastern Europe. In *Chapter Four a* multitude of both temporal and spatial scales are used to examine a range of factors that point to causes other than a lack of democracy as being responsible for environmental degradation in the Eastern European communist countries. An evaluation of environmental problems in Eastern Europe since 1991 is then made in *Chapter Five*. Conclusions arising from these chapters are then drawn in *Chapter Six*.

Chapter 1

Environmental Degradation in the Eastern European Communist Countries

The environmental problems in the communist countries of Eastern Europe were not nationwide; they were concentrated in regional and local hot spots. Some of these areas experienced extreme air pollution, high levels of nuclear risk, or exposure to hazardous waste. The Black Triangle, which covers parts of the Czech Republic, Germany and Poland, is a good example of a regional air pollution hot spot. Very high concentrations of sulphur dioxide existed in this region due to a concentration of local industry, the use of sulphur-rich brown coal, technologies which are inferior to those in the West, and unfavourable geographical conditions. It is likely that this air pollution has health implications through increasing respiratory diseases and reducing life expectancy, whilst also damaging natural features and buildings (Tellegen, 1996, p. 70).

In the case of environmental degradation through nuclear risk, the district of Chelyabinsk is a good example of a hot spot. A military complex in the area routinely dumped nuclear water in the Techa River after 1947, with Monroe estimating that 124,000 people were exposed to the radiation (Tellegen, 1996, p. 71). The Chernobyl disaster had much more widespread ecological and economic impacts, with the main impacts occurring in the Soviet Republic of Belarus, but with some impacts occurring throughout the Northern Hemisphere (Tellegen, 1996, pp. 71-2).

In terms of environmental degradation from exposure to hazardous waste, a good example of a hot spot is Schoenberg in the former GDR. Western countries have dumped colossal amounts of hazard-ous waste at the site since 1979, with over half a million tonnes coming from The Netherlands alone (Tellegen, 1996, p. 73).

Two of these 'hot spots' of environmental degradation in the communist countries of Eastern

Europe – air pollution and nuclear risk – could be argued to arise from the communist nature of those countries. Do the 'hot spots' of environmental degradation which arise from the dumping of hazardous waste into Eastern Europe by democratic states also arise from the communist nature of these countries? The waste itself is obviously *generated by* democratic states; however the waste also has to be *accepted by* the incoming country. The relationship between the creation of hazardous waste 'hot spots' and the communist nature of the Eastern European states will be explored a little later.

Chapter 2

Environmental Degradation & Communist Ideology, Economics and Politics

The communist countries of the 'Second World' were theoretically socialist countries that were passing through the 'dictatorship of the proletariat' on the way to a future communist society. There was thus a 'state capitalism' with no place for democracy to affect the allocation of productive resources. It could be argued that this framework is conducive to environmental degradation due to communist ideology, economics and politics. I will consider these three phenomena in the paragraphs which follow.

Marxist ideology was formed in an era when the overarching *global* social construction was that nature existed to be exploited by humans. According to Hunt (1962, p. 213) Russians "have always been lacking a sense of the relative, and they thus embraced it [Marxism] with an uncompromising dogmatism, so that, with the passage of time, they have converted it into an ossified and inflexible code". If Hunt is right, then this view of nature – as something which needs to be tamed, as an obstacle to progress – which is enshrined in the concept of voluntarism, would have been ingrained in the ideology which underpinned the Eastern European communist countries.

This *global* social construction was challenged in 1962 with the publication of Rachel Carson's *Silent Spring* which was the "single most effective catalyst for environmentalism" (Mcneill, 2000, p. 337). A new overarching global construction was formed concerning the interdependence of the human/nature relationship. However, reframing did not take place in Eastern Europe, and this was plausibly because of the ingrained nature of the old ideology.

In Marxist economics resources only have a use value, thus their free availability leads to inefficiency and overuse. For example, the free availability of water led to an excessive irrigation frequency of cotton crops in Uzbekistan, and also to an excessive wastage of water during transportation. This contributed significantly to the shrinking of the Aral Sea. Furthermore, central planning discourages innovation and also results in excessive resource use per product unit due to the stipulation of quantities of material inputs in planning targets.

The communist political system is also conducive to environmental degradation. It inherently prohibits the non-state countervailing power base that is required so that environmental concerns can be highlighted, and then fed into eco-friendly adjustments in productive processes. Furthermore, within the Eastern European communist governments the responsibility for the environment was integrated into ministries whose overriding objective was the economic fulfilment of planning targets.

There are clearly grounds in communist ideology, economics and politics for believing that communism promotes environmental degradation.

Chapter 3

The Industrialisation of the West

We have seen that there are components of communism that can explain environmental degradation occurring. In order to gauge the importance of these components it is useful to consider the environmental degradation that occurred when the Western democracies industrialised.

These countries had the exploitation of nature as a central component. Black (2000, p. 122) describes how:

in the mines, pregnant women were put to work hauling loads of coal. Serious injury was a constant danger; homes were filthy slums; wages were barely enough to sustain life. Death, all too often, came prematurely.

Health and life were thus subjugated to resource exploitation. It is helpful to look at examples of hot spots similar to those in Eastern Europe – severe air pollution in the UK and extreme toxic chemical dumping across the US.

The example of air pollution in the UK puts the problems of the Black Triangle into perspective. In 1873 a toxic yellow fog gripped London for three days, claiming approximately 700 lives, mainly through respiratory distress. In January 1880 another 700 people died, and the death toll continued to mount through the turn of the century (Christianson, 1999, p. 149). The severity of the problem and the scientific ignorance of air pollution in 1911 can be appreciated from the following quote from Christianson (1999, pp. 149-150):

Dr. H. A. Des Voeux, a member of the Smoke Coal Abatement Society, arranged for a more thorough analysis of the air during one such golden onslaught. The results confirmed the doctor's suspicions that in addition to the usual concentration of water and carbon dioxide, the polluted air contained deadly amounts of sulphur oxides, chlorine, ammonia, ash, and what was deemed "tarry matter"… (a) toxic brew he dubbed "smog".

Despite this knowledge the state took no action and the death toll continued to mount. It took the Great Smog of 1952 which killed 4000 people to eventually stimulate action, which was the passing of the 1956 Clean Air Act. So, although the problem was addressed, it took 83 years from the onset of the problem and 45 years from its scientific recognition. In this period industrialisation was prioritised at the expense of the environment, unimpeded by a countervailing power base.

In the case of toxic chemical dumping in the US the example of Love Canal is particularly pertinent. The response of the state to their citizen's fears that they had been poisoned is reminiscent of what one might expect from an Eastern European communist state. There were:

a series of denials from local government... (an) unwillingness of the population to believe that industry and government would allow its citizens to be poisoned...(and) trying to get 'facts' from state officials was all but impossible.

(Maples, 2003, pp. 218-219)

To get the health effects they were subjected to recognised the residents had to set up a non-state countervailing power base and fight politically through demonstrations, letters to papers and publicity campaigns. It also emerged that there are likely to be tens of thousands of similar toxic chemical hot spots across the US, where citizens are being left to be poisoned. These are a potent legacy of industrialisation (Maples, 2003, p. 220). Thus serious degradation occurred, but distributive justice through political fighting was eventually obtained.

These examples of the severe environmental degradation which occurred when the West industrialised are important ones. These examples are but the tip of the iceberg – just two examples of the massive amount of environmental degradation that occurred when these democratic states industrialised.

Chapter 4

The Range of Factors Influencing Environmental Degradation in the Eastern European Communist Countries

We have seen that there was serious environmental degradation when the democracies of the West industrialised, and that there were serious health impacts. This implies that there are factors other than communism underlying the environmental degradation in Eastern Europe.

The first factor of importance is the geographic history of the area which 'environmentally disadvantaged' the Eastern European countries – these countries were endowed with highly polluting brown coal; so, if the same amount of coal was used in Eastern Europe and the West the environmental degradation would, *ceteris paribus*, be far greater in Eastern Europe.

The other factors arise in more recent times, a primary one being industrialisation. Regardless of political regime, industrialisation causes environmental degradation, as has been seen with the examples in the previous chapter from the industrialisation of the West. So, the high levels of air pollution in Eastern European communist countries could be argued to be mainly caused by the energy-intensive stage of industrialisation that the countries were in, a stage that the West had already passed through.

The other primary factor, which is closely inter-linked to industrialisation, is the existence of the 'Cold War' between 1945 and 1991. The nuclear stalemate meant that a full-scale international war was too dangerous to be contemplated. Thus the communist strategy for winning the 'Cold War' was as follows:

The Soviet Union and the other communist-related countries would simply outproduce the capitalist nations, developing the inherently superior capacity of a socialised economy to expand production and raise standards of living. When after a few years the masses throughout the world saw that the prosperity of the Soviet Union far exceeded that of the United States, they would be irresistibly attracted to Communism and their governments would be unable to refuse their demands for imitation of the Soviet model.

(Hudson, 1968, pp. 175-176)

So, winning the 'Cold War' was the obvious overriding priority for the communist countries, and victory in this 'War' demanded a more rapid and extensive industrialisation than that which occurred in the West. This was not a slow 'natural' industrialisation, but a win-at-all-costs war driven industrialisation. This caused severe environmental degradation, due not to a lack of democracy, but due to war. The combination of the intrinsic dynamism of capitalism, combined with the lack of innovation in communism, meant that the communist countries struggled in vain to supersede Western living standards, thus increasing industrialisation pressures even further.

'Cold War' posturing was also a motivating force behind big dam projects and other big scale construction schemes, both in Eastern Europe and in the West, rather than being a solitary outcome of voluntarism. As Mcneill (2000, p. 341) asserts: "the international system, in Darwinian language, selected rigorously against ecological prudence in favour of policies dictated by short term security considerations".

Therefore, it was the international situation that dictated the extreme nature of the industrialisation in the communist countries of Eastern Europe. It should be noted that the communist states inherently did take environmental protection seriously, even though effective implementation might have been hindered by the 'Cold War'. The Soviet Union was one of the first countries to formulate environmental quality standards, which were stricter than those later formulated in the US (Tellegen, 1996, p. 80). Furthermore, in 1989 Poland had 18 per cent of its country environmentally protected; this is five times more than the area protected in 1980 (Tellegen, 1996, p. 69).

Capitalism in the West also caused environmental degradation in the Eastern European communist countries due to the motivations of western firms to minimise their costs through the legal and illegal transport of hazardous waste. It could be argued that the lack of democracy in Eastern European communist countries contributed to this environmental degradation. However, the fact that the transfer has mushroomed since democratization – with 72 foreign firms, disposers and brokers engaged in 64 waste trade schemes with 13 countries (Tellegen, 1996, p. 74) – indicates that the lack of democracy under communism helped to inhibit the capitalist offloading of hazardous waste.

Chapter 5

Environmental Degradation in Eastern European Countries since 1991

What has happened to the environment of the Eastern European countries since democracy replaced communism in 1991? Has the environment improved? Or, contrarily, has there been an increase in environmental degradation?

There have been a plethora of increases in environmental degradation in these countries since 1991. We have already seen, in the previous chapter, that the Eastern European countries have become inundated with hazardous waste *originating from capitalist countries* since 1991.

Another source of recent degradation has been the desire of the Eastern European countries to attain the living standards of the West. Many of the industries that were previously condemned as polluting were hailed as engines of economic growth after 1991 (Tellegen, 1996, p. 89). Indeed, according to Jehlicka (2004, p. 12):

in the late 1990s the overriding concern of CEE countries has become the minimisation of such (environmental) reforms on their economic growth and competitiveness.

With democracy has come the freedom to prioritise private benefits, and these benefits have adverse social costs in terms of environmental degradation. So, the environmentally friendly public transport systems of communist Eastern Europe are giving way to growing car use (Tellegen, 1996, p. 95). Indeed, after democracy came to Albania car use increased more than 13 fold, making Tirana the most polluted city in Europe. The WHO set a safe limit of 50 microgrammes per cubic metre of air for PM10s; the figure in Tirana averaged 483 mcg. This has been the major cause of a 20 per cent increase in the mortality rate in parts of Tirana in the decade following 1991 (O'Grady, 2004, p. 7).

The neoliberal reforms of waste management have also increased environmental degradation. Hungary's pre-1991 industrial waste management system which prioritised waste prevention, reuse and reduction has been replaced by an inferior system of disposal and incineration. This system has not only eroded incentives for resource conservation but also actively encouraged waste generation (Jehlicka, 2004, pp. 23-24).

Chapter 6

Conclusions

We have considered many aspects of the relationship between communism and environmental degradation. We can conclude that there are inherent components of communist ideology, politics and economics that played a part in the environmental degradation of Eastern Europe.

However, the extent of the environmental degradation that occurred when the West industrialised implies that industrialization is a major causal factor leading to environmental degradation regardless of political regime. Furthermore, the experience of the Eastern European countries since 1991 exposes both the international and the domestic pressures that cause democracies to prioritise economic growth at the expense of the environment.

We have seen in the cases of hazardous waste transfer, waste management, and transport that a lack of democracy can produce environmentally beneficial outcomes. When democracy replaced communism in Eastern Europe there was a serious increase in environmental degradation in many areas. However, as we saw in the case of Love Canal, democracy does enable citizens to fight for the environment through a non-state countervailing power base.

It is perhaps useful to distinguish between two different but related questions in order to gauge the extent to which a lack of democracy contributed to environmental degradation in the communist countries of Eastern Europe.

Firstly: *Why did the environmental degradation occur?* The answer appears to be that it occurred because of industrialisation *not* because of communism. Furthermore, this industrialisation was particularly severe/'speeded up' because of the 'Cold War' imperative and its associated posturing.

Secondly: *Why was the degradation not addressed by Eastern European governments?* Their stage of industrialisation and the existence of non-state countervailing power bases can explain why the West was able to respond effectively to the post-1962 emergence of the new human/nature paradigm. However, in Eastern Europe the 'Cold War' played a large part in preventing the implementation of widespread environmental policies; it also seems plausible that the 'ingrained' nature of the pre-1962 communist framework might also have played a 'preventative' role.

We can conclude that the lack of democracy in the Eastern European communist countries played a very minor role in causing environmental degradation, but it, and the associated international situation (the 'Cold War') can help to explain why the degradation was not addressed.

Given the early formulation of strict environmental regulations in the Soviet Union, and the multitude of environmental degradation impacts which have occurred in the Eastern European countries since 1991, there are good reasons to believe that the communist model is fundamentally compatible with a high level of environmental protection.

Bibliography

Black, J. (2000) *Encyclopedia of World History*, Bath, Parragon.

Christianson, G. (1999) *Greenhouse: The 200-year story of global warming,* London, Constable and Company Limited.

Hudson, G. F. (1968) *Fifty Years of Communism,* Middlesex, Penguin Books Ltd.

Hunt, R. N. Carew (1962) *The Theory and Practice of Communism,* Middlesex, Penguin Books Ltd.

Jehlicka, P. (2004) 'Eastern Enlargement of the European Union and the Environment' in *DU310 Updating Material: Stop Press 4, Mailing 2,* Milton Keynes, The Open University.

Maples, W. (2003) 'Environmental justice and the environmental justice movement' in Bingham, N, Blowers, A and Belshaw, C (eds) *Contested Environments,* Chichester, John Wiley & Sons/The Open University.

Mcneill, J. (2000) *Something New Under the Sun: An environmental history of the twentieth century,* London, Penguin Books.

O'Grady, J. (2004) 'Europe at a glance', *The Week,* 3 April 2004, Issue 454, p. 7.

Tellegen, E. (1996) 'Environmental conflicts in transforming economies: Central and Eastern Europe' in Sloep, P and Blowers, A (eds) *Environmental Policy in an International Context: Conflicts,* London, Arnold.

Other books by the author:

Sustainable Development & GM Food: An analysis of the relationship between the genetic modification of crops and the varieties of sustainable development (2011)

Preserving Biodiversity: The role of economics in international environmental policy-making (2011)

The Role of the Market in Environmental Protection (2011)